Contents

DIAGRAMS

Use of Guidance

THE APPROVED DOCUMENTS

This document is one of a series that has been approved by the Secretary of State as practical guidance on meeting the requirements of Schedule 1 and regulation 7 of the Building Regulations 1991 (SI 1991 No. 2768) as amended by the Building Regulations (Amendment) Regulations 1992 (SI 1992 No. 1180), the Building Regulations (Amendment) Regulations 1994 (SI 1994 No. 1850), the Building Regulations (Amendment) Regulations 1995 (SI 1995 No. 1356), the Building Regulations (Amendment) Regulations 1997 (SI 1997 No. 1904) and the Building Regulations (Amendment) Regulations 1998 (SI 1998 No. 2561).

At the back of this document is a list of those documents currently published by the Department of the Environment, Transport and the Regions and the Welsh Office which have been approved for the purpose of the Building Regulations 1991.

The Approved Documents are intended to provide guidance for some of the more common building situations. However, there may well be alternative ways of achieving compliance with the requirements. **Thus there is no obligation to adopt any particular solution contained in an Approved Document if you prefer to meet the relevant requirement in some other way.**

Other requirements

The guidance contained in an Approved Document relates only to the particular requirements of the Regulations which that document addresses. The building work will also have to comply with the requirements of any other relevant paragraphs in Schedule 1 to the Regulations.

There are Approved Documents which give guidance on each of the other requirements in Schedule 1 and on regulation 7.

LIMITATION ON REQUIREMENTS

In accordance with regulation 8, the requirements in Parts A to K and N of Schedule 1 to the Building Regulations do not require anything to be done except for the purpose of securing reasonable standards of health and safety for persons in or about the building.

MATERIALS AND WORKMANSHIP

Any building work which is subject to the requirements imposed by Schedule 1 to the Building Regulations should, in accordance with regulation 7, be carried out with proper materials and in a workmanlike manner.

You may show that you have complied with regulation 7 in a number of ways. These include the appropriate use of a product bearing CE marking in accordance with the Construction Products Directive (89/106/EEC)[1] as amended by the CE marking Directive (93/68/EEC)[2], or a product complying with an appropriate technical specification (as defined in those Directives), a British Standard, or an alternative national technical specification of any state which is a contracting party to the European Economic Area which, in use, is equivalent, or a product governed by an agrément certificate such as one issued by the British Board of Agrément. You will find further guidance in the Approved Document supporting regulation 7 on materials and workmanship.

Technical specifications

Building regulations are made for specific purposes: health and safety, energy conservation and the welfare and convenience of disabled people. Standards and technical approvals are relevant guidance to the extent that they relate to these considerations. However, they may also address other aspects of performance such as serviceability, or aspects which although they relate to health and safety are not covered by the Regulations.

When an Approved Document makes reference to a named standard, the relevant version of the standard is the one listed at the end of the publication. However, if this version of the standard has been revised or updated by the issuing standards body, the new version may be used as a source of guidance provided it continues to address the relevant requirements of the Regulations.

The Secretary of State has agreed with the British Board of Agrément on the aspects of performance which it needs to assess in preparing its Certificates in order that the Board may demonstrate the compliance of a product or system which has an Agrément Certificate with the requirements of the Regulations. An Agrément Certificate issued by the Board under these arrangements will give assurance that the product or system to which the Certificate relates, if properly used in accordance with the terms of the Certificate, will meet the relevant requirements.

Similarly, the appropriate use of a product which complies with a European Technical Approval as defined in the Construction Products Directive will also meet the relevant requirements.

---

1 As implemented by the Construction Products Regulations 1991 (SI 1991 No 1620)

2 As implemented by the Construction Products (Amendment) Regulations 1994 (SI 1994 No 3051)

Mixed use development

In mixed use developments part of a building may be used as a dwelling while another part has a non-domestic use. In such cases, if the requirements of this Part of the Regulations for dwellings and non-domestic uses differ, the requirements for non-domestic use should apply in any shared parts of the building.

The Workplace (Health, Safety and Welfare) Regulations 1992

The Workplace (Health, Safety and Welfare) Regulations 1992 contain some requirements which affect building design. The main requirements are now covered by the Building Regulations, but for further information see: *Workplace health, safety and welfare, The Workplace (Health, Safety and Welfare) Regulations 1992, Approved Code of Practice and Guidance;* The Health and Safety Commission, L24; Published by HMSO 1992; ISBN 0-11-886333-9.

The Workplace (Health, Safety and Welfare) Regulations 1992 apply to the common parts of flats and similar buildings if people such as cleaners, wardens and caretakers are employed to work in these common parts. Where the requirements of the Building Regulations that are covered by this Part do not apply to dwellings, the provisions may still be required in the situations described above in order to satisfy the Workplace Regulations.

Disability Discrimination Act 1995 and The Disability Discrimination (Employment) Regulations 1996

Following the guidance in this Approved Document is not a requirement for satisfying duties under Sections 6 and 21 of the Disability Discrimination Act 1995, to make adjustments to premises. It should be noted, however, that under Regulation 8 of the Disability Discrimination (Employment) Regulations 1996 an employer will not be required to alter any physical characteristic included within a building which was adopted with a view to satisfying the requirements of Part M of the Building Regulations and met those requirements at the time the building works were carried out and continues to substantially meet those requirements.

The Requirements

This Approved Document, which takes effect on 25 October 1999, deals with the following requirements from Part M of Schedule 1 to the Building Regulations 1991 as amended by the Building Regulations (Amendment) Regulations 1998.

Requirement	*Limits on application*
Interpretation **M1.** In this Part, 'disabled people' means people who have - (a) an impairment which limits their ability to walk or which requires them to use a wheelchair for mobility, or (b) impaired hearing or sight.	1. The requirements of this Part do not apply to - (a) a material alteration; (b) an extension to a dwelling, or any other extension which does not include a ground storey;
Access and use **M2.** Reasonable provision shall be made for disabled people to gain access to and to use the building.	(c) any part of a building which is used solely to enable the building or any service or fitting in the building to be inspected, repaired or maintained.
Sanitary conveniences **M3. (1)** Reasonable provision shall be made in the entrance storey of a dwelling for sanitary conveniences, or where the entrance storey contains no habitable rooms, reasonable provision for sanitary conveniences shall be made in either the entrance storey or a principal storey. **(2)** In this paragraph "entrance storey" means the storey which contains the principal entrance to the dwelling, and "principal storey" means the storey nearest to the entrance storey which contains a habitable room, or if there are two such storeys equally near, either such storey. **(3)** If sanitary conveniences are provided in any building which is not a dwelling, reasonable provision shall be made for disabled people.	
Audience or spectator seating **M4.** If the building contains audience or spectator seating, reasonable provision shall be made to accommodate disabled people.	2. Part M4 does not apply to dwellings.

Notes

Attention is drawn to the Disability Discrimination (Employment) Regulations 1996
Under Regulation 8 of the Disability Discrimination (Employment) Regulations 1996 an employer will not be required to alter any physical characteristic included within a building which was adopted with a view to satisfying the requirements of Part M of the Building Regulations and met those requirements at the time the building works were carried out and continues to substantially meet those requirements.

Attention is drawn to the Workplace (Health, Safety and Welfare) Regulations 1992
Compliance with Building Regulation requirement M2, in conjunction with Part K, where it relates to stairs and ramps would, in accordance with Section 23(3) of the Health and Safety at Work, etc Act 1974, prevent the service of an improvement notice with regard to the requirements of Regulation 17 of the Workplace (Health, Safety and Welfare) Regulations 1992 which relate to permanent stairs, ladders and ramps on pedestrian traffic routes within the workplace premises.

Alterations: Part M is referred to in regulation 3(2)(b) which defines 'material alterations'. Regulation 4(2) stipulates that a material alteration should not result in an altered building being less satisfactory in respect of access and facilities for disabled people than it was before.

Means of escape in case of fire: the scope of Part M is limited to matters of access to, into, and use of, a building. It does not extend to means of escape for disabled people in the event of fire, for which reference should be made to Approved Document B, Fire safety.

Stairs and ramps: Approved Document K, Protection from falling, collision and impact contains general guidance on stair and ramp design. Part M contains more specific provisions for stairs and ramps that need to be suitable for use by disabled people.

Guidance

Performance

In the Secretary of State's view the requirements of Part M will be met by making it reasonably safe and convenient for disabled people to:

a. gain access to and within, buildings other than dwellings and to use them. The provisions for access and facilities are for the benefit of disabled people who are visitors to the building or who work in it.

b. visit new dwellings and to use the principal storey. The provisions are expected to enable occupants to cope better with reducing mobility and to 'stay put' longer in their own homes, although not necessarily to facilitate fully independent living for all disabled people.

Where the requirements apply

New buildings

0.1 The requirements apply if a building is newly erected, or has been substantially demolished to leave only external walls.

0.2 If, as part of the reconstruction of a building, other than a dwelling, it is impractical to make adjustments to the level of the existing principal entrance or any other appropriate existing entrance, to permit independent access for wheelchair users, or to provide a new entrance which is suitable, the other requirements of Part M should still apply.

Extensions

0.3 If an existing building, other than a dwelling (note extensions to dwellings are excluded from Part M), is extended, the requirements of Part M apply to the extension provided that it contains a ground storey.

0.4 When a building is extended, there is no obligation to carry out improvements within the existing building to make it more accessible to and usable by disabled people than it was before. However the extension should not adversely affect the existing building with respect to the provisions of the Building Regulations for access to, and use of, the building by disabled people.

0.5 An extension should be at least as accessible to and usable by disabled people as the building being extended. Where access to the extension is achieved only through the existing building, it will be subject to the limitations of the existing building, and it would be unreasonable to require higher standards within the extension. On the other hand, it is reasonable that an extension which is independently approached and entered from the boundary of the site should be treated in the same manner as a new building.

Alterations

0.6 When a building is altered there is no obligation to improve access and facilities for disabled people. However the level of provision after alteration should not be any worse. Facilities may be moved but their suitability and access to them should not be reduced.

External features

0.7 Part M applies to those features, outside the building, which are needed to provide access to the building from the edge of the site and from car parking within the site.

What requirements apply

0.8 If Part M applies, reasonable provision should be made in:

i) Buildings other than dwellings

a. so that disabled people can reach the principal entrance to the building and other entrances described in this Approved Document, from the edge of the site curtilage and from car parking within the curtilage;

b. so that elements of the building do not constitute a hazard for a person with an impairment of sight;

c. for access for disabled persons into and within any storey of the building and to any facilities provided to comply with Part M;

d. so that disabled people can use the building's facilities;

e. for sanitary accommodation for disabled people;

f. for suitable accommodation for disabled people in audience or spectator seating; and

g. for aids to communication for people with an impairment of hearing or sight in auditoria, meeting rooms, reception areas and ticket offices.

ii) Dwellings (including any purpose-built student living accommodation, other than a traditional halls of residence providing mainly bedrooms and not equipped as self-contained accommodation)

a. so that disabled people can reach the principal, or suitable alternative, entrance to the dwelling from the point of access;

b. for access for disabled persons into and within the principal storey of the dwelling; and

c. for sanitary accommodation at no higher storey than the principal storey.

Educational establishments

0.9 In schools or other educational establishments, Requirements M2, M3(3) and M4 will be satisfied if the provisions comply with paragraphs 2.1/2/4/6, 3.1, 4.1/2/4/6 and 5.1 in Design Note 18, 1984 'Access for Disabled Persons to Educational Buildings', published by the Secretary of State for Education and Employment (or the relevant parts of any update of the Design Note). The introduction of provisions for people with impaired hearing or sight means that Requirement M2 may need to be satisfied by incorporating into these buildings some of the features described in Design Note 18 as general design considerations.

Definitions

0.10 The following meanings apply to terms throughout this Approved Document.

Access, approach or entry.

Accessible, with respect to buildings or parts of buildings, means that disabled people are able to gain access.

Suitable, with respect to means of access and facilities, means that they are designed for use by disabled people.

Principal entrance storey, the storey which contains the principal entrance or entrances to the building. If an alternative accessible entrance is to be provided by virtue of paragraph 1.31(b), the storey containing that entrance is the principal entrance storey.

Building, in this Approved Document, means a building or a part of a building which may comprise individual premises: a shop, an office, a factory, a warehouse, a school or other educational establishment including student living accommodation in traditional halls of residence, an institution, or any premises to which the public is admitted whether on immediate payment, fee, subscription, or otherwise.

0.11 The following meanings apply only to terms used in the sections on dwellings in this Approved Document.

Common, serving more than one dwelling.

Habitable room, for the purpose of defining the principal storey, means a room used, or intended to be used, for dwelling purposes, including a kitchen but not a bathroom or a utility room.

Maisonette, a self-contained dwelling, but not a dwelling-house, which occupies more than one storey in a building.

Point of access, the point at which a person visiting a dwelling would normally alight from a vehicle which may be within or outside the plot, prior to approaching the dwelling.

Principal entrance, the entrance which a visitor not familiar with the dwelling would normally expect to approach or the common entrance to a block of flats.

Plot gradient, the gradient measured between the finished floor level of the dwelling and the point of access.

Steeply sloping plot, a plot gradient of more than 1 in 15.

Section 1

MEANS OF ACCESS TO AND INTO BUILDINGS OTHER THAN DWELLINGS

Objectives

1.1 The aim is to provide a suitable means of access for disabled people to the building from the point of entrance to the site curtilage and from car parking which is provided within the building site. It is also important that external circulation which is proposed between different parts of the building is suitable.

1.2 In designing the approach to an entrance, it should be recognised that wheelchair users and ambulant disabled people have difficulty in negotiating changes of level. People with impaired sight may be unaware of the onset of abrupt changes in level.

1.3 The design of the approach will also need to take account of overall constraints on space.

1.4 It is important to reduce the risks to disabled people, particularly those with impaired sight, of being injured when passing close to the building. This means that parts of the building should not present hazards on circulation routes immediately adjacent to it.

1.5 Disabled people should be able to use the principal entrance provided for visitors or customers and an entrance which is intended, exclusively, for members of staff.

1.6 The needs of disabled people vary. Alternative means of access are helpful. Not all ambulant disabled people find it as easy to negotiate a ramp as they do a stair.

'Level' approach from the edge of site and car parking

Design considerations

1.7 Gradients should be as gentle as the circumstances allow. They will then be more convenient for wheelchair users and other people with walking difficulties. Where possible, a 'level' approach should be provided.

1.8 People who use wheelchairs, stick or crutches or who are blind or partially sighted, and those who may accompany them, need adequate space when approaching the building. There should also be space for people passing in the opposite direction.

Provisions

1.9 A 'level approach' will satisfy Requirement M2 if it has a surface width of at least 1.2m and its gradient is not steeper than 1 in 20.

1.10 If site constraints necessitate an approach steeper than 1 in 20, a ramped approach should be provided.

1.11 Where a pedestrian route to the building is intended for use by disabled people:

a. a tactile warning should be provided for people with impaired vision where the route crosses a carriageway and at the top of steps; and

b. dropped kerbs should be provided for wheelchair users.

Diagram 2 illustrates paving slabs with tactile warning services. The blister type paving is considered suitable for use at pedestrian crossing points. The corduroy type paving is considered suitable for use at the top of external stairs.

Ramped approach

Design Considerations

1.12 It may not always be possible to arrange a 'level' approach. Where a ramped approach is necessary, the gradient should still be as gentle as possible. Steep gradients create difficulties for some wheelchair users who lack the strength to propel themselves up a slope or have difficulty in slowing down or stopping when descending. Nor are they as safe or convenient for ambulant people with disabilities. As well as adding to problems of unsteadiness in adverse weather conditions, they increase the risk of slipping.

1.13 Some disabled people, or their helpers, need to be able to stop frequently: for instance, to regain strength or breathe, or to ease pain.

1.14 Wheelchair users need adequate space to stop on landings, to open and to pass through doors without the need to reverse into circulation routes or to face the risk of rolling back down slopes.

1.15 Design considerations for the width of ramped approaches are similar to those for level approaches.

1.16 Some disabled people have a weakness on one side or the other and that leads to the need for support at each side of ramped approaches.

1.17 The risk of wheelchair users catching their feet beneath or between balustrade rails should be minimised. This can be achieved by providing kerbs or solid balustrades on open sides.

1.18 Where practicable, easy going steps should complement ramped approaches.

Provisions

1.19 A ramped approach will satisfy Requirement M2 if it:

a. has a surface which reduces the risk of slipping;

b. has flights whose surface widths are at least 1.2m and whose unobstructed widths are at least 1.0m;

c. is not steeper than 1 in 15, if individual flights are not longer than 10.0m, or not steeper than 1 in 12, if individual flights are not longer than 5.0m;

d. has top and bottom landings, each of whose lengths is not less than 1.2m and, if necessary, intermediate landings, each of whose lengths is not less than 1.5m, in all cases clear of any door swing;

e. has a raised kerb at least 100mm high on any open side of a flight or a landing; and

f. has a continuous suitable handrail on each side of flights and landings, if the length of the ramp exceeds 2.0m.

Diagram 1 illustrates the guidance on ramped approaches.

Diagram 1 Ramped approach with complementary steps

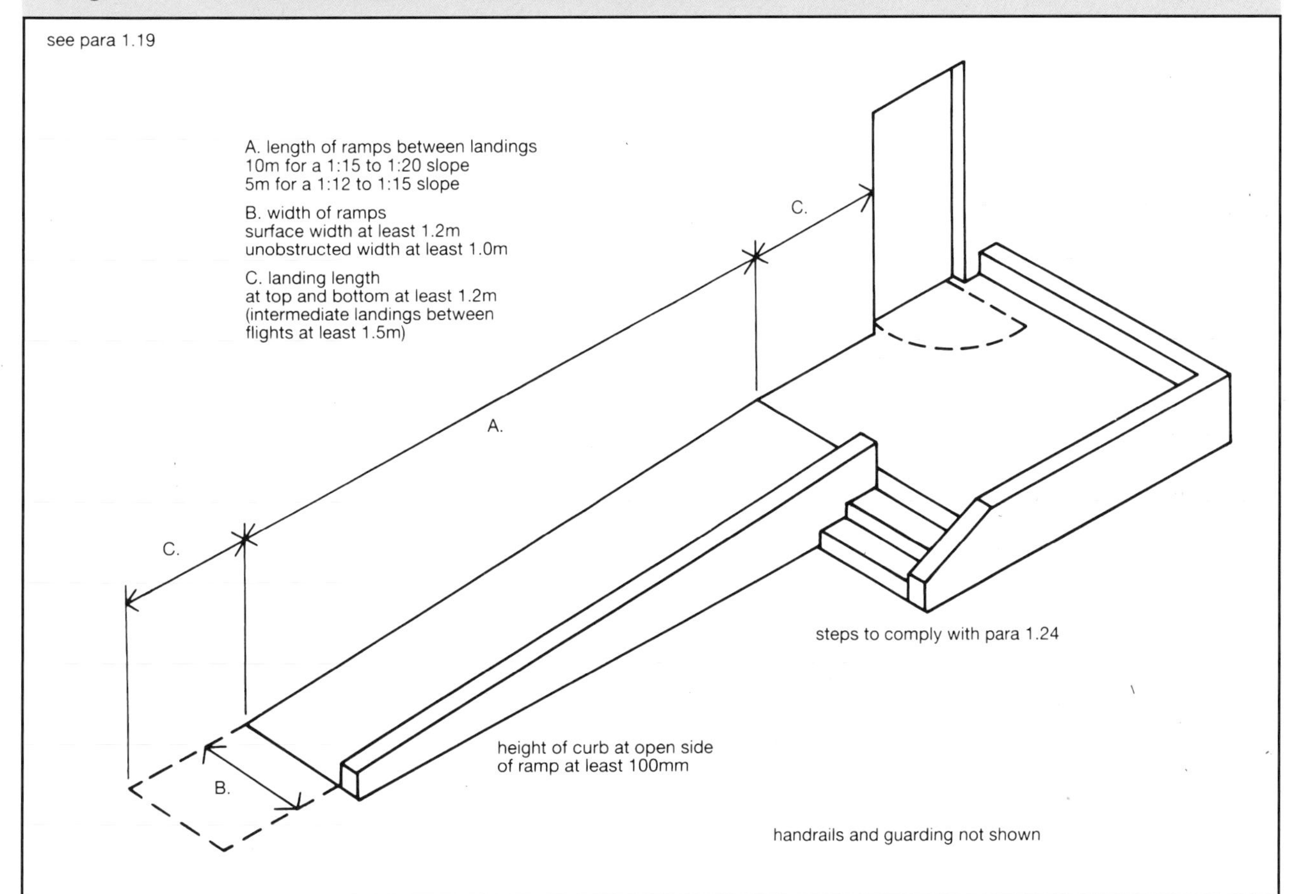

Stepped Approach

Design considerations

1.20 Whilst similar design considerations apply to stepped as do to ramped approaches, there are additional ones.

1.21 People with impaired sight are at risk of tripping or losing their balance when meeting sudden changes of level. The danger is at its greatest when approaching the head of a flight of steps. The existence of individual steps, on their own or within a flight, should also be made apparent.

1.22 People who wear callipers or who have stiffness in hip or knee joints are particularly at risk of tripping or catching their feet beneath nosings or treads. Physical weakness on one side or the other and sight impairments necessitate tread dimensions which allow both feet to be placed square onto it.

1.23 These design considerations apply to a stepped approach which is provided to satisfy the objectives in paragraphs 1.1 - 1.6.

Provisions

1.24 A stepped approach will satisfy Requirement M2 if:

a. its top landing has a tactile surface, to give advance warning of the change in level;

b. all step nosings are distinguishable through contrasting brightness;

c. it has flights whose unobstructed widths are at least 1.0m;

d. the rise of a flight between landings is not more than 1.2m;

e. it has top and bottom and, if necessary, intermediate landings, each of whose lengths is not less than 1.2m clear of any door swing onto it;

f. the rise of each step is uniform and not more than 150mm;

g. the going of each step is not less than 280mm, which for tapered treads should be measured at a point 270mm from the 'inside' of the stair;

h. risers are not open; and

j. there is a suitable continuous handrail on each side of the flight and landings if the rise of the stepped approach comprises two or more risers.

Diagram 2 shows corduroy paving that will provide a tactile warning at the top of the steps. Diagram 3 shows the location of visual and tactile warnings. Diagram 4 illustrates the guidance on stepped approaches.

Diagram 2 **Tactile paving slabs**

see para 1.11 & 1.24

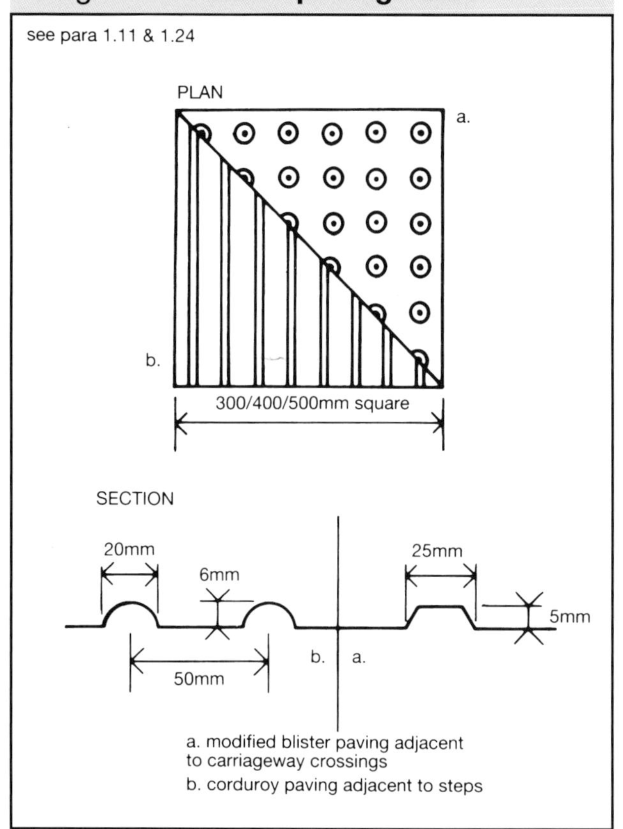

Diagram 3 **Tactile and visual warnings**

see para 1.24

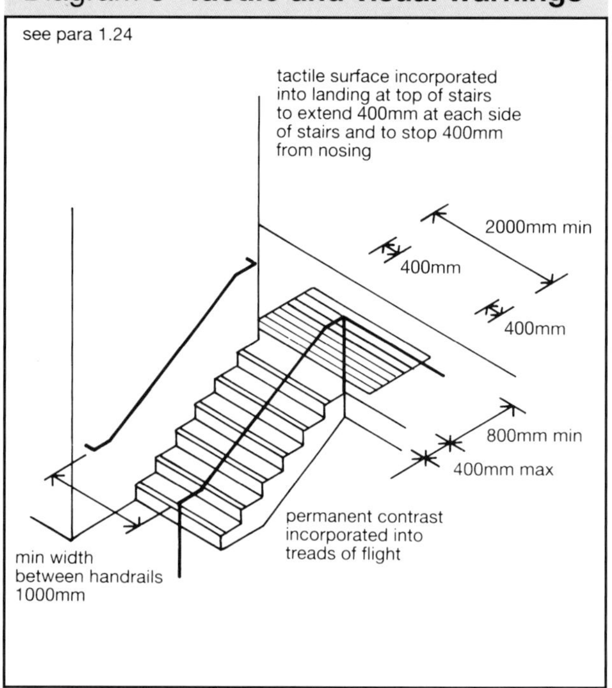

Diagram 4 **Stepped approach**

see para 1.24

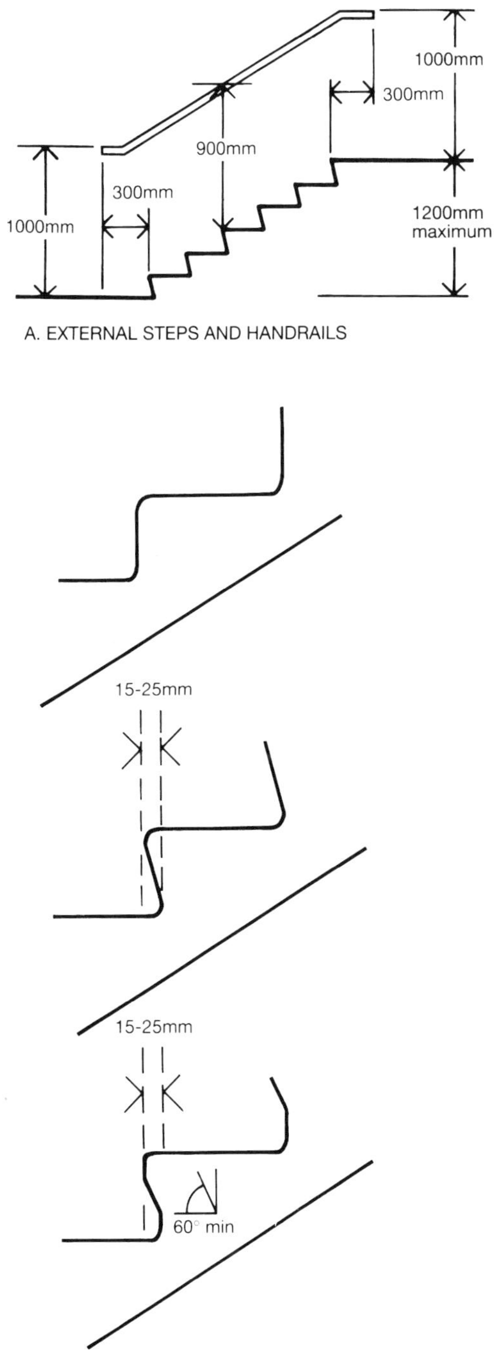

1000mm
300mm
900mm
300mm
1000mm
1200mm maximum

A. EXTERNAL STEPS AND HANDRAILS

15-25mm
15-25mm
60° min

B. EXAMPLES OF SUITABLE TREAD NOSING PROFILES
MAXIMUM RISE 150mm
MINIMUM GOING 280mm

Handrails

Design consideration

1.25 For those who have physical difficulty in negotiating changes of level, grippable and well supported handrails are important.

Provisions

1.26 Requirement M2 will be satisfied if:

a. the top of a handrail is 900mm above the surface of a ramp or the pitch line of a flight of steps and 1000mm above the surface of a landing;

b. the handrail extends at least 300mm beyond the top and bottom of a ramp, or the top and bottom nosings of a stepped approach, and terminates in a closed end which does not project into a route of travel; and

c. the profile of the handrail and its projection from a wall is suitable.

Diagram 5 contains guidance on a handrail design that would satisfy Requirement M2.

Diagram 5 **Handrail design**

see para 1.26

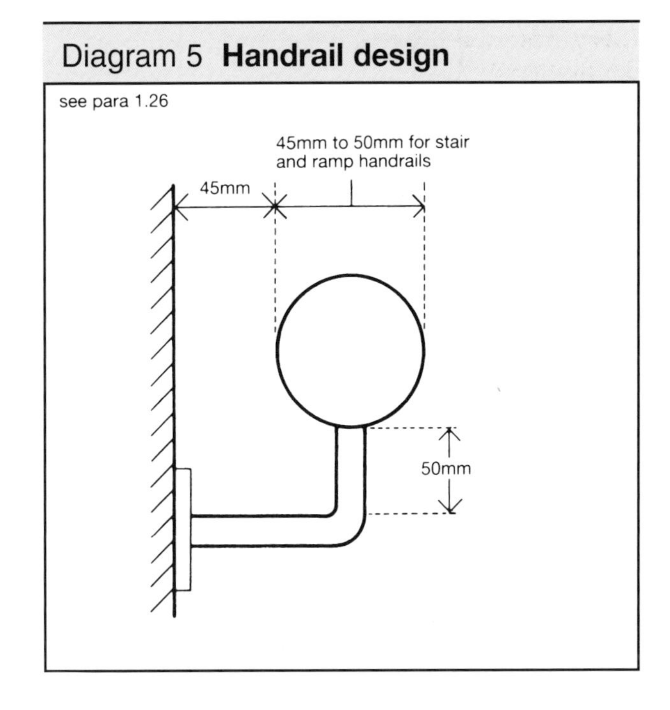

Hazards on access routes

Design considerations

1.27 Features of the building that occasionally obstruct a route adjacent to the building may be a hazard to people with sight impairments, particularly if the object is partially transparent and therefore indistinct.

Provisions

1.28 Windows or doors in general use which open outwards should not cause an obstruction on a path which runs along the face of a building.

1.29 Diagram 6 contains guidance on the reduction of these risks.

Diagram 6 **External hazards**

see para 1.28

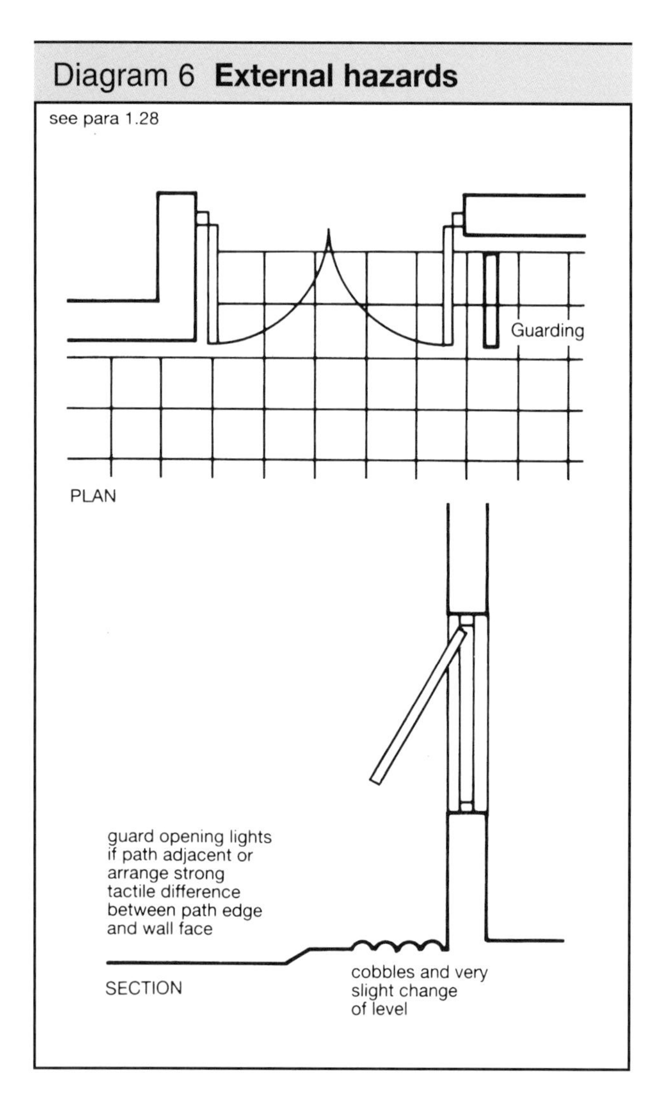

Access into the building

Design considerations

1.30 There should be a convenient access into the building for disabled people whether they are visitors to the building or work in it and whether they arrive on foot or in a wheelchair.

Provisions

1.31 Requirement M2 will be satisfied if:

a. the principal entrance for visitors or customers is accessible and suitable;

b. in the event of the space outside the principal entrance being severely restricted, or the site being on sloping ground, an alternative entrance intended for general use is accessible and suitable and there is suitable internal access, available to all who may use the building, from the alternative entrance to the principal entrance;

c. where car spaces are provided adjacent to and serving the building, but there is no suitable means of access from them to the principal entrance, an additional entrance, intended for general use, is provided giving suitable internal access to the principal entrance; and

d. an entrance which is provided specifically for members of staff is accessible and suitable.

Principal entrance doors

Design considerations

1.32 Sufficient width should be available for wheelchair manoeuvre. The opportunity should be taken for more generous planning than might be available internally.

1.33 A space provided alongside the leading edge of a door reduces the risk of a wheelchair user being prevented from reaching the door handle as a result of the wheelchair footrest colliding with the return wall.

1.34 People with mobility difficulties cannot react quickly to avoid collisions and, where feasible, they should be able to see people approaching the other side of entrances and should, themselves, be seen.

Provisions

1.35 Requirement M2 will be satisfied if a principal entrance door:

a. contains a leaf which provides a minimum clear opening width of not less than 800mm;

b. has an unobstructed space on the side next to the leading edge for at least 300mm, (unless the door is opened by a suitable automatic control); and

c. is provided with a glazed panel giving a zone of visibility from a height of 900mm to 1500mm from the finished floor level wherever the opening action of the door could constitute a hazard.

Diagram 7 illustrates the guidance on doors.

Note: Clear opening widths in excess of 800mm can be achieved by selecting a 1000mm single leaf doorset (850mm clear opening width of leaf) or a 1.8m double leaf doorset (810mm clear opening width of each leaf) - as shown in Table 2 to BS 4787: *Internal and external wood doorsets, door leaves and frames* Part 1: 1980(1985) *Specification for dimensional requirements.*

Diagram 7 **Entrance doorways**

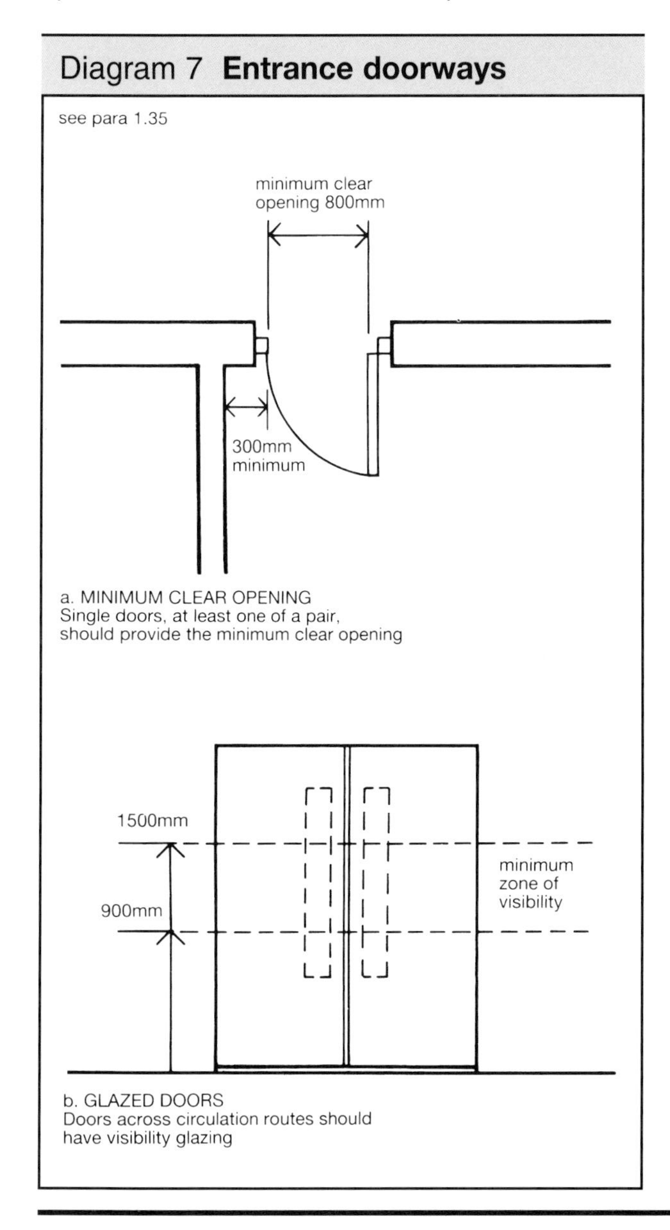

Revolving doors

Design considerations

1.36 By themselves, small revolving doors are not suitable for use by disabled people. Timing of entry and exit may create difficulties for people with sight impairments or with ambulatory problems. In addition, there may be insufficient space within the confines of the door to accommodate a wheelchair or there may be too little time for manoeuvre.

1.37 Some larger types of revolving doors are considered suitable on their own. They should be capable of accommodating several people at the same time. They revolve very slowly and are equipped with mechanisms to slow them down further and to stop them as soon as they feel resistance.

Provisions

1.38 An entrance fitted with a small revolving door would be suitable if it also contained a door as described in paragraph 1.35 and Diagram 7.

1.39 Some large types of revolving door are considered suitable on their own (see paragraph 1.37).

Entrance lobbies

Design considerations

1.40 A wheelchair user should be able to move clear of one door before using the next one. There should also be space for someone assisting the wheelchair user and for someone passing in the opposite direction.

Provisions

1.41 Requirement M2 will be satisfied if an entrance lobby is designed on the basis of the examples shown in Diagram 8.

Diagram 8 **Entrance lobbies**

see para 1.41

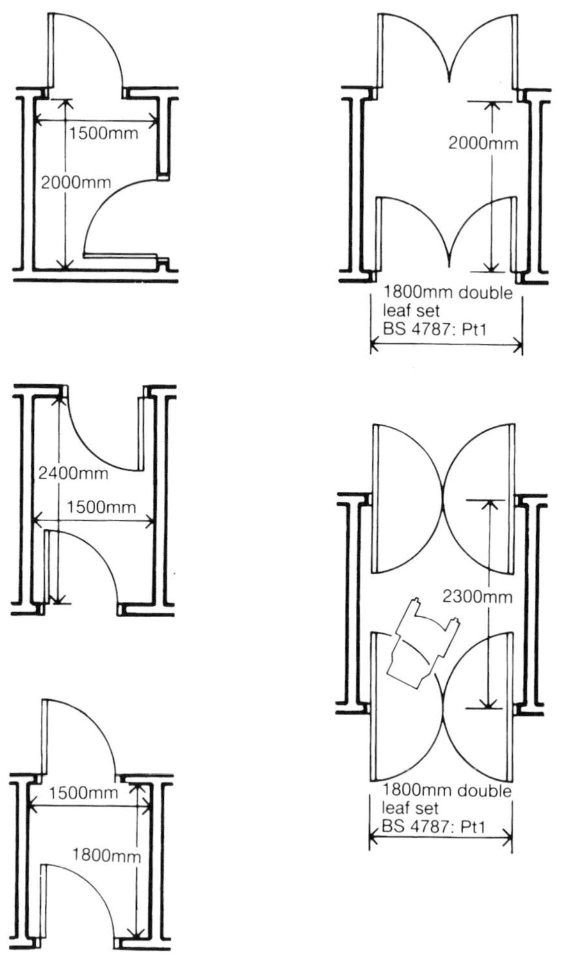

Section 2

MEANS OF ACCESS WITHIN BUILDINGS OTHER THAN DWELLINGS

Objective

2.1 The objective is similar to that relating to the approach to buildings. The need is to facilitate movement within buildings. Much has to do with the provision of sufficient space for wheelchair manoeuvre, convenient ways of travelling from one storey to another and the inclusion of features which will help those with impaired hearing or sight to find their way safely and conveniently.

2.2 Whilst the guidance included in this Approved Document is not focused entirely on the needs of wheelchair users, the more generous space criteria often relate to the space required to manoeuvre a wheelchair.

Horizontal circulation within the building

Internal doors

Design considerations

2.3 Considerations, similar to those set out in paragraphs 1.32 to 1.34 apply to the design of internal doors.

Provisions

2.4 Requirement M2 will be satisfied if:

a. an internal door contains a leaf which provides a minimum clear opening of not less than 750mm.

b. the space into which the door opens is unobstructed on the side next to the leading edge for at least 300mm, unless the door can be opened by an automatic control, or is in a situation where it may be reasonable to anticipate assistance, e.g. when leaving a fellow guest's hotel bedroom.

c. each doorway across an accessible corridor or passageway is provided with a glazed panel, giving a zone of visibility from a height of 900mm to 1500mm from the finished floor level.

Note: Clear opening widths in excess of 750mm can be achieved by using a 900mm single leaf doorset (770mm clear opening width of leaf) or one leaf of an 1800mm double leaf doorset (820mm clear opening width of door leaf) - as shown in Table 1 of BS 4787: Part 1.

Diagram 9 **Internal doorways**

see para 2.4

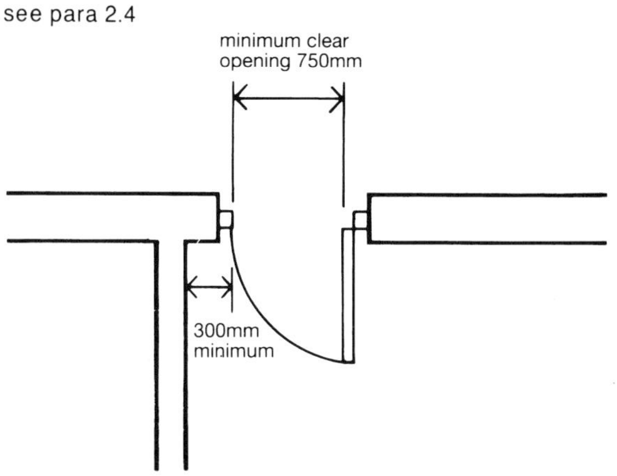

minimum clear opening 750mm

300mm minimum

a. MINIMUM CLEAR OPENING
Single doors, at least one of a pair, should provide the minimum clear opening

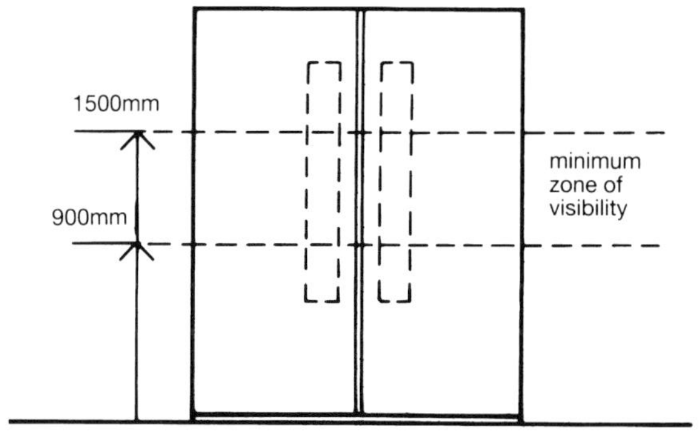

1500mm

900mm

minimum zone of visibility

b. GLAZED DOORS
Doors across circulation routes should have visibility glazing

Corridors and passageways

Design considerations

2.5 In locations required to be accessible to wheelchair users, corridors and passageways need to be wide enough to allow for wheelchair manoeuvre and for other people to pass. Narrower corridors would be reasonable in other locations, such as those to which lift access is not provided or in some extensions.

Provisions

2.6 Requirement M2 will be satisfied if a corridor or passageway:

a. to which wheelchair users have access, has an unobstructed width of 1200mm; or

b. which is accessible by stairway alone or is in an extension approached through an existing building has an unobstructed width of at least 1000mm.

Internal lobbies

Design considerations

2.7 Internal lobbies are less likely to be in demand by several people at the same time. It is therefore reasonable to adopt less generous space standards than for principal entrance lobbies. Nevertheless, a wheelchair user should be able to move clear of one door before using the next one. Provided these practical considerations are met, smaller internal lobbies may be used and will impose less constraint on the internal planning of the building.

Provisions

2.8 Requirement M2 will be satisfied if an internal lobby is designed on the basis of the examples shown in Diagram 10.

Vertical circulation within the building

Design considerations

2.9 The most suitable means of access for disabled people when passing from one storey to another is a passenger lift. However, given the added cost and intrusion into usable space it is not reasonable to require a lift to be provided in every instance.

Diagram 10 **Internal lobbies**

see para 2.8

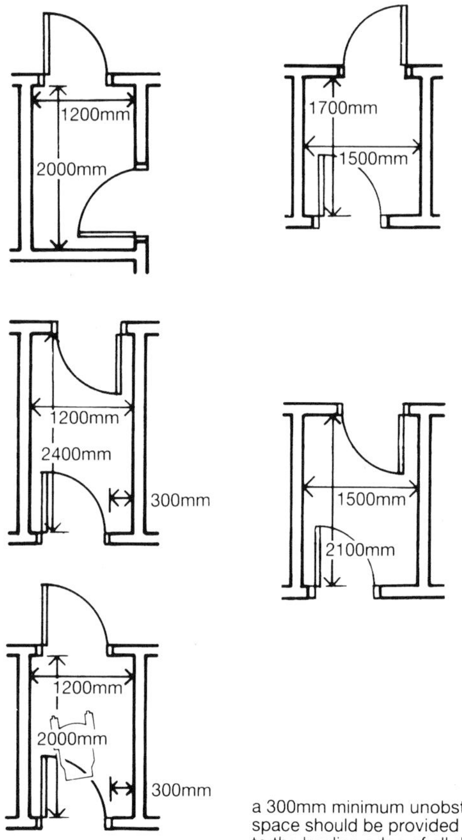

a 300mm minimum unobstructed space should be provided next to the leading edge of all doors

2.10 If there is no passenger lift to provide vertical access, a stair should be designed to satisfy the needs of ambulant disabled people. In any event, a stair should be designed to be suitable for people with impaired sight.

2.11 It would be reasonable to base the decision with regard to the provision of mechanical means of vertical access on the nett floor area of the storey to be reached.

Passenger lifts

Design considerations

2.12 A wheelchair user needs sufficient space and time to manoeuvre into a lift and, once in, should not be restricted for space.

He or she should also be able to reach the controls which summon and direct the lift. People with sensory impairments should, in some circumstances, be advised of the floor that the lift has reached. Measures should be adopted which give a disabled person time to enter the lift to reduce the likelihood of contact with closing doors.

Provisions

2.13 Requirement M2 will be satisfied if a suitable passenger lift is provided to serve any storey above or below the principal entrance storey, and that storey contains nett floor areas as follows:

a. in a two storey building, more than 280m^2 of nett floor area; or

b. in a building of more than two storeys, more than 200m^2 of nett floor area; and

c. a suitable means of access is provided from the lift to the remainder of the storey.

Note: The floor area of a storey in a building covered by the Requirement should be measured as follows:-

The area of all parts of a storey which use the same entrance from the street or an indoor mall should be added together, whether they are in more than one part of the same storey, or used for different purposes. The area of any vertical circulation, sanitary accommodation and maintenance areas in the storey should not be included.

2.14 Requirement M2 will be satisfied if a passenger lift:

a. has a clear landing at least 1500mm wide and at least 1500mm long in front of its entrance;

b. has a door or doors which provide a clear opening width of at least 800mm;

c. has a car whose width is at least 1100mm and whose length is at least 1400mm;

d. has landing and car controls which are not less than 900mm and not more than 1200mm above the landing and the car floor at a distance of at least 400mm from the front wall;

e. is accompanied by suitable tactile indication on the landing and adjacent to the lift call button to identify the storey in question;

f. which serves more than three floors, is provided with suitable tactile indication on or adjacent to lift buttons within the car to confirm the door selected;

Diagram 11 **Lift dimensions**

see para 2.14

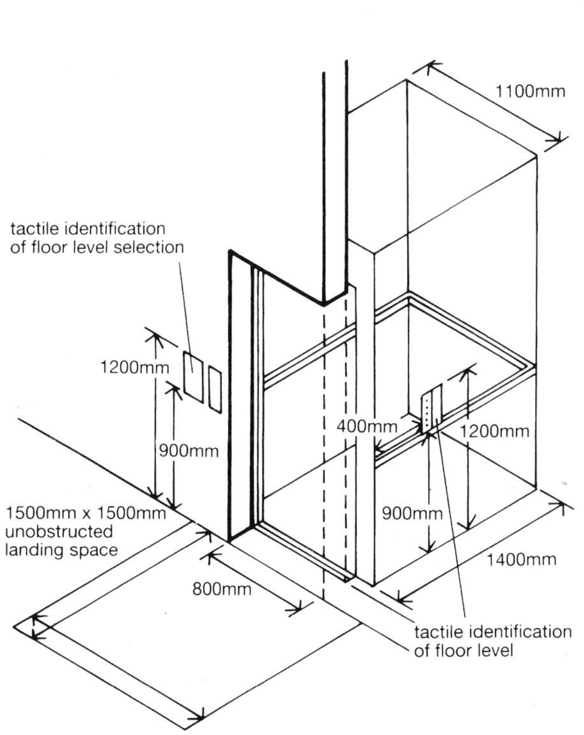

g. which serves more than three storeys, is provided with visual indication and with voice indication of the floor reached; and

h. incorporates a signalling system which gives 5 seconds notification that the lift is answering a landing call and a 'dwell time' of 5 seconds before its doors begin to close after they are fully open: the system may be overridden by a door re-activating device which relies on photo-eye or infra-red methods, but not a door edge pressure system, provided that the minimum time for a lift door to remain fully open is 3 seconds.

Diagram 11 illustrates a suitable passenger lift.

Note: Details of some of these provisions are contained in BS 5655: Parts 1 and 2 also in Part 5: 1989 *Specification for dimensions of standard lift arrangements* and Part 7: 1983 *Specification for manual control devices, indicators and additional fittings*. It is a prerequisite of BS 5655 that automatic doors to passenger lifts should be equipped with re-opening activators. These may be operated through invisible beam or contact with the person.

Wheelchair stairlifts

Design considerations

2.15 In a building containing small areas with a unique function, it may be reasonable to expect access for wheelchair users to upper and lower storeys but be impractical to provide a passenger lift. In such circumstances, a wheelchair stairlift to BS 5776: 1996 *Specification for powered stairlifts* would constitute a reasonable alternative.

A unique facility which anyone using the building should reasonably expect to use may consist, for instance, of a small library gallery, a staff rest room or a training room. In the absence of a practical alternative, it would be reasonable to install a wheelchair stairlift.

Provisions

2.16 If a storey, with a nett floor area exceeding $100m^2$, contains a unique facility but is not large enough to warrant passenger lift access described in paragraph 2.14, it should be accessible to wheelchair users.

Platform lifts

Design considerations

2.17 The provision of a ramp to effect a change in level within a storey would be reasonable, but may have serious planning implications. The problem could be mitigated by using a platform lift: though not at the expense of a stair for use by ambulant people.

Provisions

2.18 Requirement M2 will be satisfied by installing a platform lift if it is impractical to effect a ramped change in level within a storey accessible to wheelchair users. Such provision should complement stair access to effect the same change in level. Guidance relating to the design of platform lifts is contained in BS 6440: 1983 *Powered lifting platforms for use by disabled people.*

Internal stairs

Design considerations

2.19 The design considerations for internal stairs in a building in which a lift is not provided are similar to those for stepped approaches. However, design constraints are likely to be more onerous for internal stairs and they may need to be constructed at steeper pitches and to have less frequent landings. For internal stairs it is not considered reasonable to require the provision of tactile warnings of the onset of changes of level. The nosing of each stair should, however, be distinguishable for the benefit of people with impaired vision.

2.20 If the is no lift access in a building, a stair suitable for people with walking difficulties should be provided. In any event, a stair should be suitable for people with impaired sight.

Provisions

2.21 An internal stair will satisfy Requirement M2 if:

a. it has flights whose unobstructed widths are at least 1000mm;

b. all step nosings are distinguishable through contrasting brightness;

c. the rise of a flight between landings is not more than 1800mm;

d. it has top and bottom and, if necessary, intermediate landings, each of whose lengths is not less than 1200mm clear of any door swing into it;

e. the rise of each step is uniform and not more than 170mm;

f. the going of each step is uniform and not less than 250mm, which for tapered treads should be measured at a point 270mm from the 'inside' of the stair;

g. risers are not open; and

h. there is a suitable continuous handrail on each side of flights and landings if the rise of the stair comprises two or more risers.

2.22 Exceptionally, the provisions of the rise of a flight may be varied if particular storey heights of the need to gain access beneath an intermediate landing dictate, or if the additional length of the stair has unreasonable effects on usable floor areas. In such cases it would be reasonable to provide the number of risers which would satisfy the requirement K1 (see Approved Document K, Stairs, ramps and guards).

Diagram 12 illustrates the guidance on internal stairs.

Diagram 12 Internal stairs

see para 2.21

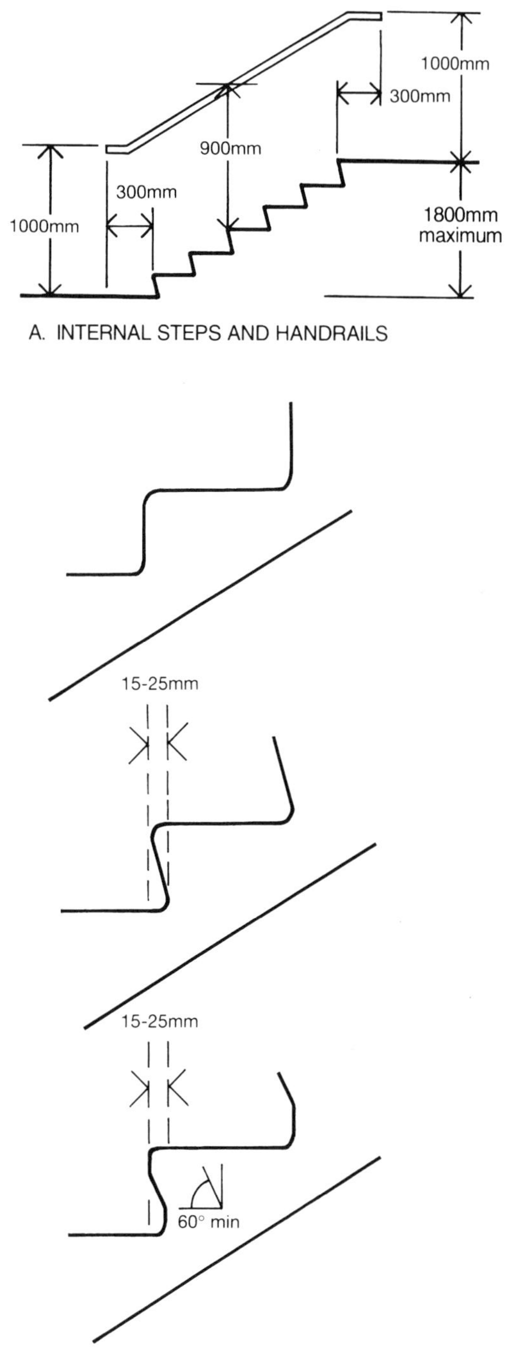

A. INTERNAL STEPS AND HANDRAILS

B. EXAMPLES OF SUITABLE TREAD NOSING PROFILES
Maximum rise 170mm
Minimum going 250mm

Internal ramps

Design considerations

2.23 The design considerations for internal ramps are similar to those for ramped approaches.

Provisions

2.24 Requirement M2 will be satisfied if a ramp complies with the provisions contained in paragraph 1.19.

Section 3

USE OF BUILDINGS OTHER THAN DWELLINGS

Objectives

3.1 In designing buildings, it is important that disabled people are able to reach the facilities that are provided within them and to use them.

3.2 Different types of buildings contain unique facilities. It is not the aim of this Approved Document to include an exhaustive list of such facilities and to provide guidance on them. The scale and nature of such things as sanitary accommodation and the provision of wheelchair space in theatres and auditoria are, anyway, subject to other requirements within Part M. Nevertheless, the opportunity has been taken to include guidance on a few facilities. One important aim is to enable people with hearing impairments to be able to play a full part in conferences, committee meetings and the like.

3.3 Common facilities, e.g. canteens and cloakrooms, doctors' and dentists' consulting rooms or other health facilities, should be located in a storey to which wheelchair users have access.

Restaurants and bars

Design considerations

3.4 The design of restaurants and bars should reflect the fact that disabled people should be able to visit them, independently or with companions. It is important that bars and self-service counters are accessible and that there is suitable access from them to seating areas. Where premises contain both self- service and waiter service, it would be reasonable for disabled people to have access to both.

3.5 Changes of floor level within seating areas create difficulties for some disabled people. Nevertheless, it would be unreasonable to regulate against all changes in level which may be intended to increase the visual impact of interior design: provided that they can be kept to a reasonable scale and they remain accessible to ambulant disabled people.

Provisions

3.6 Requirement M2 will be satisfied if:

a. suitable access is available to the full range of services offered;

b. all bars and self service counters and at least half the area where seating is provided are accessible to wheelchair users; and

c. in circumstances in which the nature of the services varies and as a result is divided into different areas in the same or different storeys, at least half of each area is accessible to wheelchair users.

Hotel and motel bedrooms

Design considerations

3.7 Disabled people who use wheelchairs need a bedroom which is accessible and is sufficiently spacious and arranged to allow manoeuvring of a wheelchair within it and into an 'en suite' bathroom, if provided.

3.8 Wheelchair users may need to gain access to bedrooms other than their own when, for instance, attending conferences or on holiday with their families. Bearing in mind the need to conserve space, it may be reasonable in these instances to limit the provision to that needed to pass through the doors to those rooms and to assume that the other guest will open and close them.

Provisions

3.9 Requirement M2 will be satisfied if:

a. one guest bedroom out of every twenty or part thereof of guest bedrooms, is suitable in terms of size, layout and facilities for use by a person who uses a wheelchair;

b. the entrance door to a guest bedroom which is designed for use by a person in a wheelchair complies with the guidance in paragraph 2.4(a) and (b); and

c. the entrance door to any other guest bedroom has a clear opening width of 750mm but with the option to dispense with the 300mm space at the side of the door.

Diagram 13 is a single example of an accessible hotel bedroom.

Diagram 13 **One example of an 'accessible' hotel bedroom and en suite bathroom**

see para 3.9

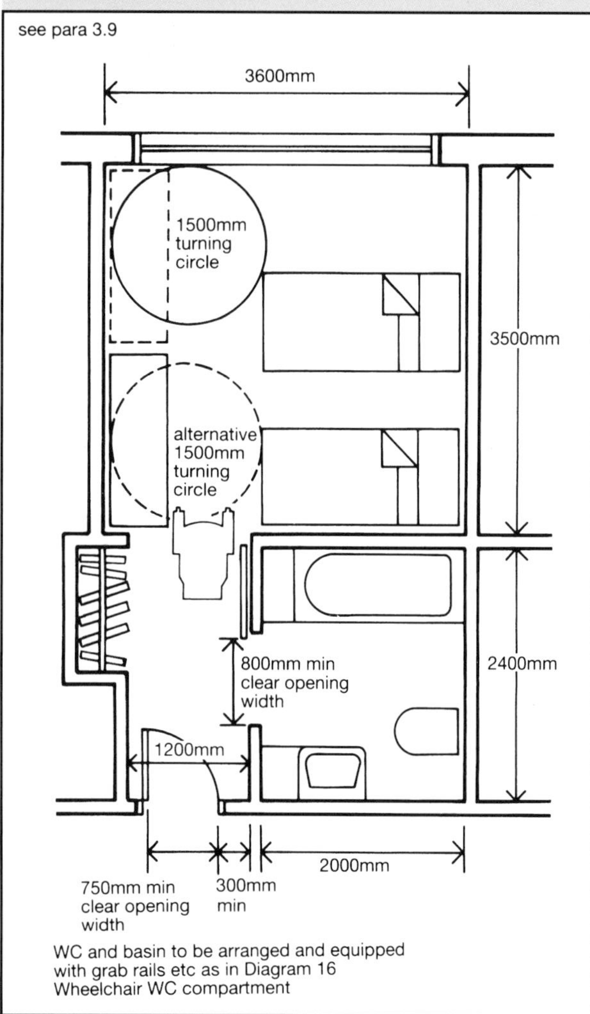

Changing facilities

Design considerations

3.10 One of the features which militates against people with disabilities taking part in recreational activities is the absence of suitable changing facilities. The provision of manoeuvring space for a wheelchair, for transfer onto a seat, and of seats, taps, shower heads, mirrors and clothes hooks mounted at suitable heights, are all critical issues.

Provisions

3.11 Requirement M2 will be satisfied if changing rooms in swimming pools and other recreational buildings contain the facilities shown in Diagrams 14 and 15.

Diagram 14 **Shower compartment**

see para 3.11

shower curtain rail
2200mm
shower to be variable between 1.2m to 2.2m from floor level
1400mm
1200mm
fixed grab rails
900mm
900mm
700mm
area of shower controls
475mm
tip up seat
SECTION
900mm
slope on floor to drain
1400mm
fixed grab rails
unobstructed approach
variable height shower
1000mm
tip up seat
shower curtain
1200mm
900mm
PLAN

THE BUILDING REGULATIONS 1991 APPROVED DOCUMENT M

Access and Facilities for Disabled People

ISBN 0 11 753469 2

The Building Regulations (Amendment) Regulations 1998, which amend Part M and the supporting Approved Document, come into force on 25 October 1999, but some conditions apply on and after 1 June 1999. The amendment extends Part M to apply to new housing.

Full details of the transitional arrangements are included in SI 1998 No. 2561, which is available from The Stationery Office (ISBN 0-11-079717-5, price £1.10). For convenience, the transitional arrangements are summarised below:

- the new requirements in Part M will not apply to a house or other relevant building already under construction on 25 October 1999, provided that the construction work began in accordance with a building notice/deposit of plans and a commencement notice, or in accordance with an initial notice, amendment notice or public body's notice;

- houses and other relevant buildings where construction work starts on or after 25 October 1999 will in general have to comply with the new requirements in Part M; but

- where full plans of a building have been deposited with and passed by a local authority before 1 June 1999, or have been the subject of a plans certificate given by an approved inspector before 1 June 1999, and accepted by the local authority either before or after that date, the new requirements in Part M will not apply to the erection of that building even if it starts on or after 25 October 1999.

The 1999 edition of the Approved Document for Part M gives guidance for both housing and non-domestic buildings. For non-domestic buildings, the guidance in Sections 1-5 of the 1999 edition is, for practical purposes, the same as that given in the 1992 edition.

Department of the Environment, Transport and the Regions
Welsh Office

February 1999

London: The Stationery Office

Diagram 15 **Dressing cubicle**

see para 3.11

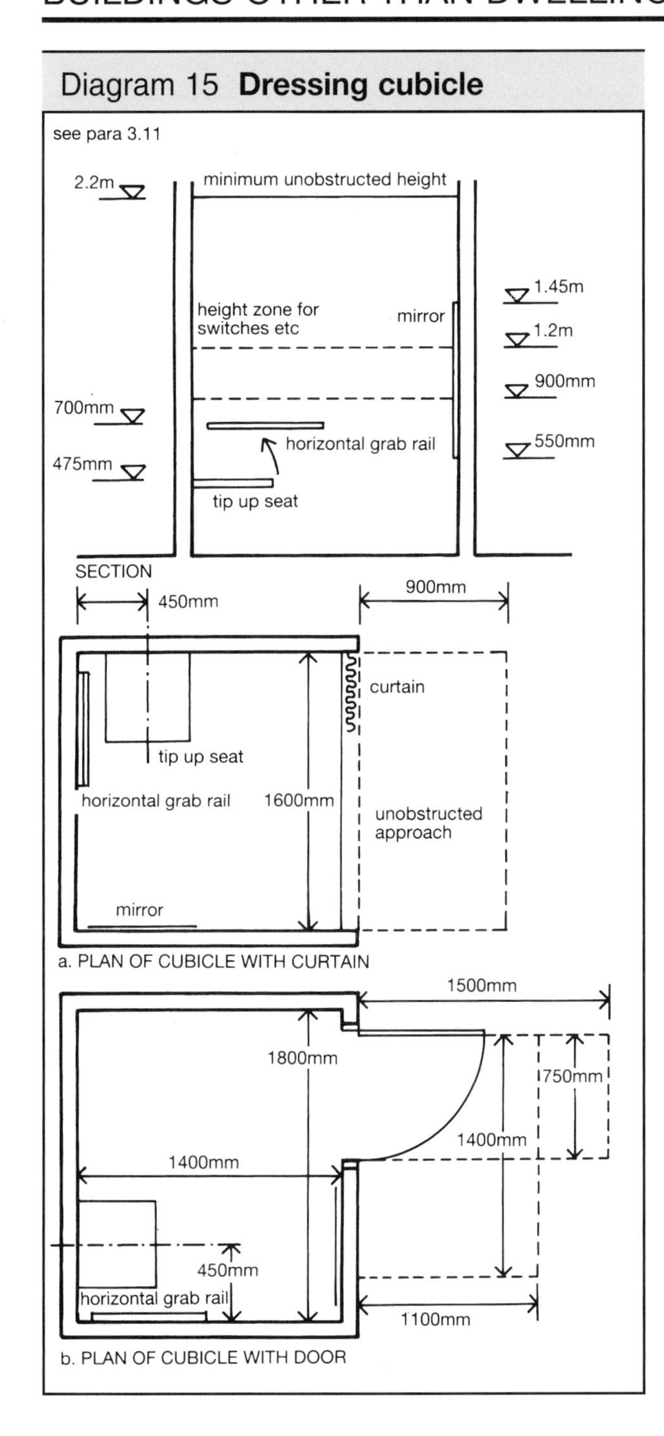

a. PLAN OF CUBICLE WITH CURTAIN

b. PLAN OF CUBICLE WITH DOOR

Aids to communication

Design considerations

3.12 In order to have the full benefit of attending a public performance or playing a proper part in discussions, a person with impaired hearing needs to receive a signal some 20dB above that received by a person with 'normal' hearing. Whichever system is selected it should also be able to suppress reverberation, and audience and other environmental noise.

3.13 The two systems most commonly used are loop induction and infra-red systems. The former depends on a signal from a microphone being passed to an amplifier which directs a current through a loop around the relevant space. A magnetic field which is generated is picked up by a listener's hearing aid and is converted into familiar sound. The infra-red system radiates invisible light which is picked up by a personal receiver, demodulated and converted into familiar sound.

3.14 A loop induction system may allow sound to spill beyond the boundary of the loop, and therefore, for those that need it, confidentiality is more difficult to achieve. That possibility is far more remote with the infra-red system, but the listener is required to wear a stethoscope for reception.

Provisions

3.15 Requirement M2 will be satisfied if aids to communication are provided at booking and ticket offices where the customer is separated from the vendor by a glazed screen and in large reception areas, in auditoria and meeting rooms in excess of 100m^2 in area.

3.16 Such systems should incorporate features which afford to a person wearing a hearing aid, the benefit of receiving sound without loss or distortion though bad acoustics or extraneous noise.

3.17 It is for the building owner to decide which system better suits the layout and use of the building and to plan accordingly.

Section 4

SANITARY CONVENIENCES IN BUILDINGS OTHER THAN DWELLINGS

Objectives

4.1 In principle, sanitary conveniences should be no less available for disabled people than for able-bodied people. The aim is provide solutions which will most reasonably satisfy that principle, whilst bearing in mind the nature and scale of the building in which the provision is to be made.

Design considerations

4.2 Some disabled people need to get to a WC quickly. Travel distances should reflect that fact.

4.3 The number and location of WCs for disabled people may depend on the size of the building and on the ease of access to the facility. A wheelchair user should not have to travel more than one storey to reach a suitable WC.

4.4 The design of the WC compartments should reflect ease of access and use at any time.

4.5 Sanitary accommodation for wheelchair users can be provided on a 'unisex' or 'integral' basis.

a. A 'unisex' facility is approached separately from other sanitary accommodation. It has practical advantages: it is more easily identified, it is more likely to be available when needed and it permits assistance by a companion of either sex. Overall, it is less demanding of space than 'integral' provision which would have to be duplicated to achieve the same level of provision for both sexes.

b. An 'integral' facility is contained within the traditional separate provision for men and women. Existing custom would preclude assistance from a member of the opposite sex to that for whom the provision is made.

To achieve flexibility, a mixture of each can be provided if the building is large enough to warrant several.

4.6 Whether the WC compartments for wheelchair users are designed on a 'unisex' or 'integral' basis, they should be similar in layout and content and should satisfy the needs:

a. to achieve necessary wheelchair manoeuvre;

b. to allow for frontal, lateral, diagonal and backward transfer onto the WC and to have facilities for hand washing and hand drying within reach from the WC, prior to transfer back onto the wheelchair; and

c. to have space to allow a helper to assist in the transfer.

4.7 Where sanitary accommodation is to be provided in upper or lower storeys without lift access, the aim should be to make reasonable provision for people who are unsteady on their feet or who need some support to stand up or sit down.

4.8 Different considerations apply to sanitary accommodation for disabled people who work in a building from that provided for disabled visitors and customers. Someone in employment might be less likely to need assistance than a visitor or a customer. Where assistance is needed by an employed person it is more likely that it will be provided by a person of the same sex, whereas a disabled visitor or customer is more likely to be accompanied by a companion of the opposite sex.

Provisions for wheelchair users

Visitors and customers

4.9 Requirement M3 will be satisfied if sanitary conveniences provided for use by visitors and customers consist of 'unisex' compartments.

Hotel and motel guest bedrooms

4.10 Requirement M3 will be satisfied if:

a. suitable 'en suite' sanitary accommodation is included with those guest bedrooms which are designed to be suitable for a disabled person, where that is the arrangement for the rest of the bedrooms; or

b. 'unisex' sanitary accommodation is provided nearby, if the general sanitary arrangement for guest bedrooms is not 'en suite'.

4.11 These facilities for hotel or motel guests are in addition to those provided in other locations in the premises by virtue of paragraphs 4.9 and 4.13 to 4.17.

4.12 Sanitary accommodation for visitors and customers, including that provided by virtue of paragraphs 4.10 and 4.11 will be suitable if designed in accordance with Diagram 16.

Staff

4.13 Requirement M3 will be satisfied if WC provision for disabled people is 'integral' within the traditional separate facilities for men and women, or is 'unisex'.

4.14 Requirement M3 will be satisfied by provision for wheelchair users of both sexes on alternate floors: provided that the cumulative horizontal travel distances from a work station to the WC is not more than 40m and, in a building provided with lift access, the general provision for sanitary conveniences is in areas to which anyone using the building has unrestricted access.

4.15 In a building which has stair access only, suitable sanitary accommodation for wheelchair users should be provided in the principal entrance storey unless that storey contains only the principal entrance and vertical circulation areas.

4.16 A WC suitable for wheelchair users should have at least the dimensions, equipment and fittings shown in Diagram 16.

4.17 If a building contains more than one WC compartment for wheelchair users, the opportunity should be taken of providing both left-hand and right-hand transfer layouts.

Diagram 16 **Wheelchair WC compartment**

see para 4.16

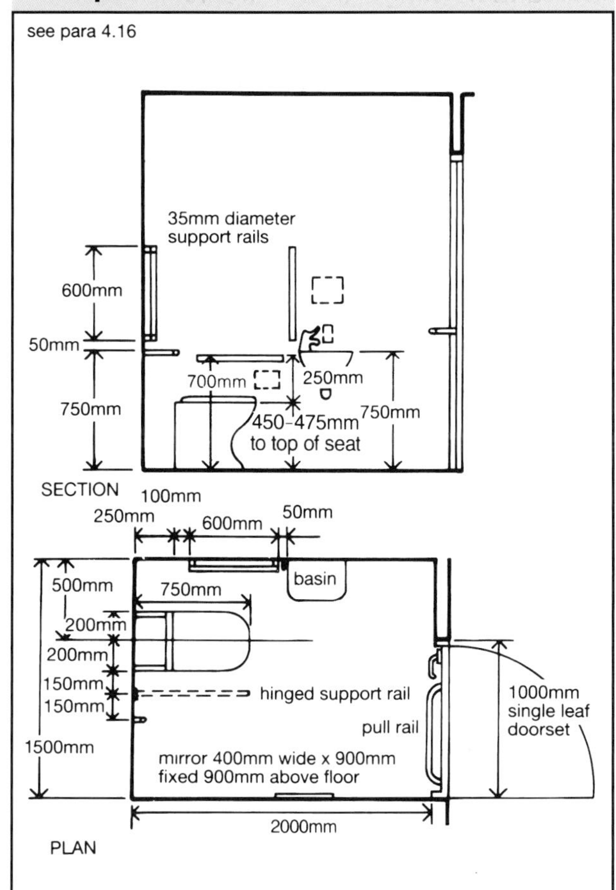

Provisions for ambulant disabled people

4.18 Requirement M3 will be satisfied if some provision, in storeys to which the only access is by a stairway, is suitable for people with a limited ability to walk and to support themselves.

4.19 At least one WC compartment designed for ambulant disabled people should be provided within each range of WC compartments included in storeys which are not designed to be accessible to wheelchair users. This is in addition to provision included under paragraph 4.15.

4.20 Diagram 17 illustrates a WC for ambulant disabled people.

Diagram 17 **WC compartment for ambulant disabled people**

see para 4.20

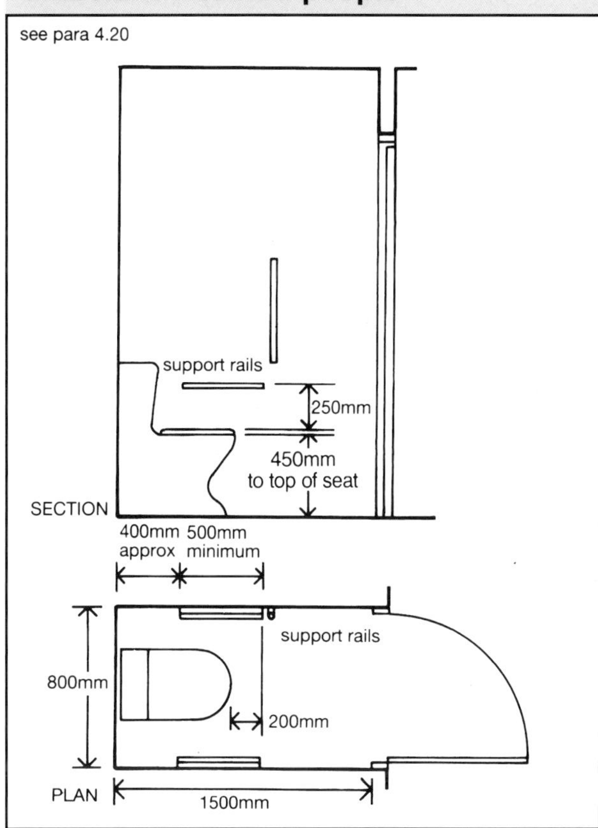

Section 5

AUDIENCE OR SPECTATOR SEATING IN BUILDINGS OTHER THAN DWELLINGS

Objectives

5.1 The aims are to make reasonable provisions for wheelchair users in theatres, cinemas, concert halls, sports stadia and the like.

Design considerations

5.2 Wheelchair users need to be provided with a space into which they can manoeuvre easily and which allows them a clear view of the event. In addition, they should have the choice of being able to sit next to disabled or able-bodied companions.

Provisions

5.3 The requirements of Part M will be satisfied by the provisions in paragraphs 5.4 - 5.6.

5.4 Of the total of fixed audience or spectator seats available to the public 6 or 1/100th, whichever is greater, should be 'wheelchair spaces'. In a large stadium it would be reasonable to provide a smaller proportion of wheelchair spaces.

5.5 In a theatre, 'wheelchair spaces' are located in a similar manner to that shown in Diagram 18. In a stadium, 'wheelchair spaces' are designed in a similar manner to those shown in Diagram 19.

5.6 A 'wheelchair space' can be provided by a clear space with a width of at least 900mm and a depth of at least 1400mm, accessible to a wheelchair user and providing a clear view of the event. The space may be one which is kept clear or be one which can readily be provided for the occasion by removing a seat. These spaces should be dispersed among the remainder of the places so that disabled people may sit next to able-bodied or disabled companions.

NOTE: Guidance on access for disabled people to sports stadia is included in the following:

Guide to Safety at Sports Grounds

Designing for spectators with disabilities

Access for disabled people

Diagram 18 **Disposition of wheelchair spaces in a theatre**

see para 5.5 & 5.6

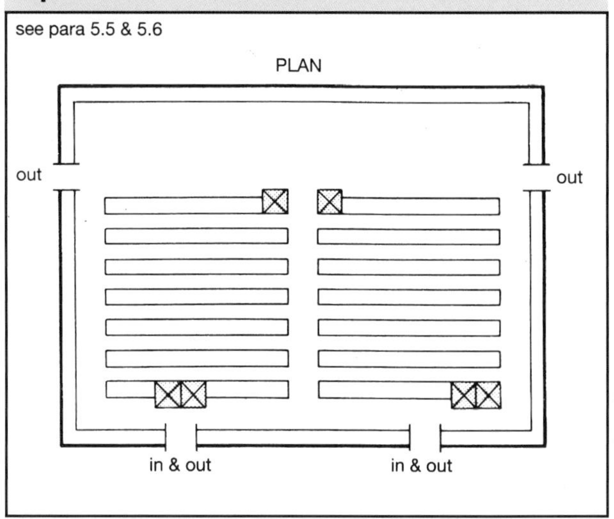

Diagram 19 **Viewing positions for disabled people in a stadium area**

see para 5.5

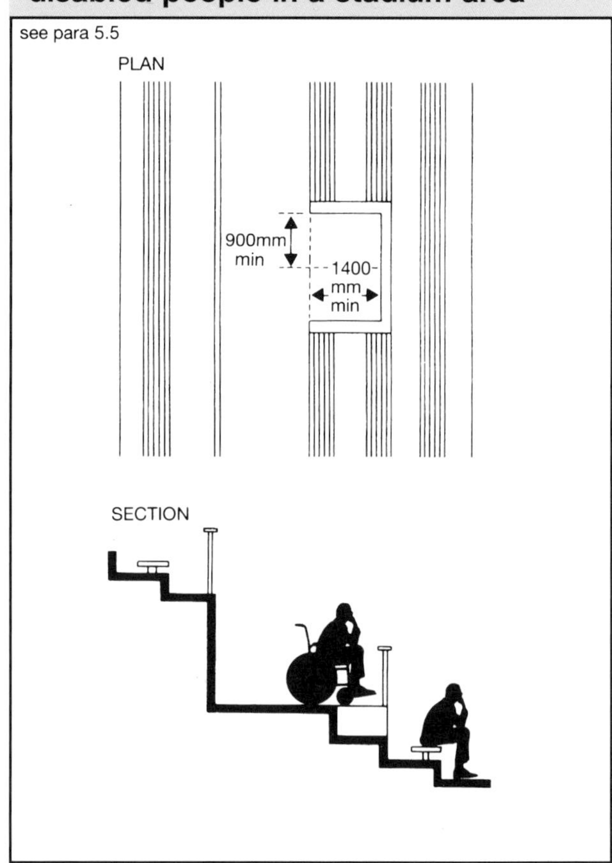

Section 6

MEANS OF ACCESS TO AND INTO THE DWELLING

Objective

6.1 The objective is to make reasonable provision within the boundary of the plot of the dwelling for a disabled person to approach and gain access into the dwelling from the point of alighting from a vehicle which may be within or outside the plot. In most circumstances it should be possible to provide a level or ramped approach.

6.2 On plots which are reasonably level, wheelchair users should normally be able to approach the principal entrance. Exceptionally, for more steeply sloping plots, it is considered reasonable to provide for stick or crutch users (see paragraph 6.9).

6.3 On plots where wheelchair users have approached the entrance, they should also be able to gain access into the dwelling-house and entrance level flats.

APPROACH TO THE DWELLING

Design considerations

6.4 The provision of an approach which can be used by disabled people, including wheelchair users, will often be a matter of practicability. Variations in topography, available plot area, or the distance of the dwelling from the point of access, may all influence the type of approach that can be provided.

6.5 Normally, the provisions will apply to the approach to the principal entrance. However, if that is not possible in a particular situation, it would be reasonable to apply them to the approach to a suitable alternative entrance.

6.6 The approach should be as safe and as convenient for disabled people as is reasonable and, ideally, be level or ramped. However, on steeply sloping plots a stepped approach would be reasonable

6.7 If a stepped approach to the dwelling is unavoidable, the aim should be for the steps to be designed to suit the needs of ambulant disabled people (see paragraph 6.17).

6.8 Alternatively, the presence of a driveway might provide a better opportunity for creating a level or ramped approach, particularly if it also provides the sole means of approach for visitors who are disabled. The driveway itself could be designed as the approach from the pavement or footpath or may be the place where visitors park. In such cases, a level or ramped approach may be possible from the car parking space, particularly on steeply sloping plots.

6.9 It is important that the surface of an approach available to a wheelchair user should be firm enough to support the weight of the user and his or her wheelchair and smooth enough to permit easy manoeuvre. It should also take account of the needs of stick and crutch users. Loose laid materials, such as gravel or shingle, are unsuitable for the approach.

6.10 The width of the approach, excluding space for parked vehicles, should take account of the needs of a wheelchair user, or a stick or crutch user (see paragraph 6.13).

NOTE:
Account will also need to be taken of planning requirements, such as for new building within conservation areas. Location and arrangement of dwellings on the site is a matter for planning, whereas the internal layout and construction of the dwellings is a matter for building control.

Provisions

6.11 The Requirement will be satisfied if, within the plot of the dwelling, a suitable approach is provided from the point of access to the entrance. The point of access should be reasonably level and the approach should not have crossfalls greater than 1 in 40.

6.12 The whole, or part, of the approach may be a driveway.

Level approach

6.13 A 'level' approach will satisfy the Requirement if its gradient is not steeper than 1 in 20, its surface is firm and even and its width is not less than 900mm.

Ramped approach

6.14 If the topography is such that the route from the point of access towards the entrance has a plot gradient exceeding 1 in 20 but not exceeding 1 in 15, the Requirement will be satisfied if a ramped approach is provided.

6.15 A ramped approach will satisfy the Requirement if it:

a. has a surface which is firm and even;

b. has flights whose unobstructed widths are at least 900mm;

c. has individual flights not longer than 10.0m for gradients not steeper than 1 in 15, or 5.0m for gradients not steeper than 1 in 12; and

d. has top and bottom landings and, if necessary, intermediate landings, each of whose lengths is not less than 1.2m, exclusive of the swing of any door or gate which opens onto it.

Stepped approach

6.16 If the topography is such that the route (see paragraphs 6.6-6.8) from the point of access to the entrance has a plot gradient exceeding 1 in 15, the Requirement will be satisfied if a stepped approach is provided.

6.17 A stepped approach will satisfy the Requirement if:

a. it has flights whose unobstructed widths are at least 900mm;

b. the rise of a flight between landings is not more than 1.8m;

c. it has top and bottom and, if necessary, intermediate landings, each of whose lengths is not less than 900mm;

d. it has steps with suitable tread nosing profiles (see Diagram 20) and the rise of each step is uniform and is between 75mm and 150mm;

e. the going of each step is not less than 280mm, which for tapered treads should be measured at a point 270mm from the 'inside' of the tread; and

f. where the flight comprises three or more risers, there is a suitable continuous handrail on one side of the flight. A suitable handrail should have a grippable profile; be between 850mm and 1000mm above the pitch line of the flight; and extend 300mm beyond the top and bottom nosings.

Diagram 20 **External step profiles**

see para 6.17

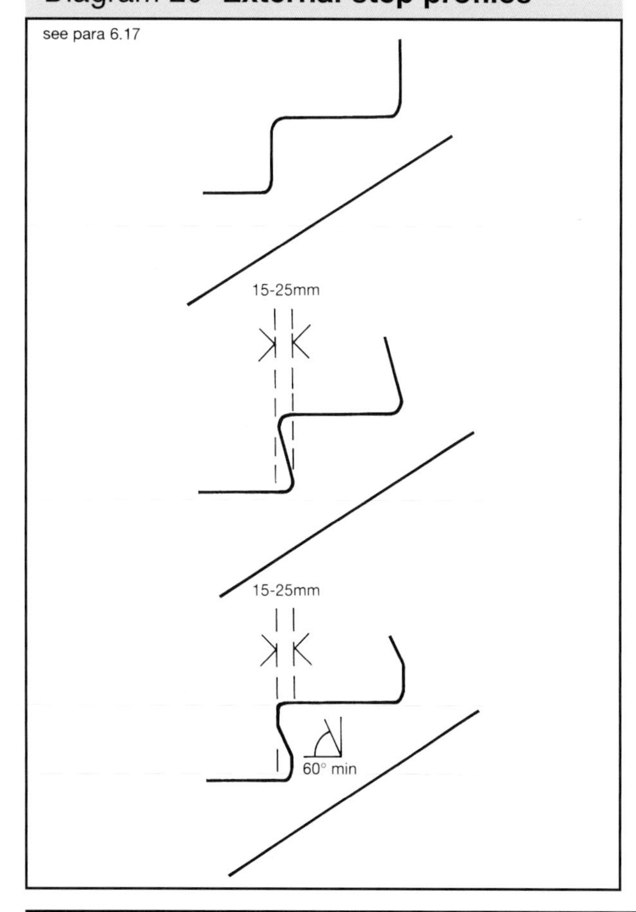

Approach using a driveway

6.18 Where a driveway provides a means of approach towards the entrance, it will satisfy the Requirement if the driveway provides an approach past any parked cars in accordance with paragraphs 6.11-6.17, above.

ACCESS INTO THE DWELLING

Design considerations

6.19 Where the approach to the entrance consists of a level or ramped approach (see paragraphs 6.13-6.15), an accessible threshold at the entrance should be provided. An accessible threshold into entrance level flats should also be provided.

6.20 In exceptional circumstances where the approach to the entrance consists of a stepped approach (see paragraph 6.16), it would still be reasonable to provide an accessible threshold. If a step into the dwelling is unavoidable, the rise should be no more than 150mm.

Provisions

6.21 If the approach to the dwelling or block of flats consists of a level or ramped approach, the Requirement will be satisfied if an accessible threshold is provided into the entrance. The design of an accessible threshold should also satisfy the requirements of Part C2: *Dangerous and offensive substances* and Part C4: *Resistance to weather and ground moisture.*

NOTE: General guidance on design considerations for accessible thresholds will be published separately.

Entrance doors

Design considerations

6.22 The provision of an appropriate door opening width will enable a wheelchair user to manoeuvre into the dwelling.

Provisions

6.23 The Requirement will be satisfied if an external door providing access for disabled people has a minimum clear opening width of 775mm.

Section 7

CIRCULATION WITHIN THE ENTRANCE STOREY OF THE DWELLING

Objective

7.1 The objective is to facilitate access within the entrance storey or the principal storey of the dwelling, into habitable rooms and a room containing a WC, which may be a bathroom on that level.

Corridors, passageways and internal doors within the entrance storey

Design considerations

7.2 Corridors and passageways in the entrance storey should be sufficiently wide to allow convenient circulation by a wheelchair user. Consideration should be given to the effects of local obstruction by radiators and other fixtures.

7.3 It will be necessary to consider the layout of a room served by an alternative to the principal entrance in order that a wheelchair user can pass through it to reach the remainder of the entrance storey.

7.4 Internal doors need to be of a suitable width to facilitate wheelchair manoeuvre. A wider door than generally provided would allow easier manoeuvring when it is necessary for a wheelchair user to turn into a door opening, as opposed to approaching it head-on.

Provisions

7.5 The Requirement will be satisfied if:

a. a corridor or other access route in the entrance storey or principal storey serving habitable rooms and a room containing a WC (which may be a bathroom) on that level, has an unobstructed width in accordance with Table 1;

b. a short length (no more than 2m) of local permanent obstruction in a corridor, such as a radiator, would be acceptable provided that the unobstructed width of the corridor is not less than 750mm for that length, and the local permanent obstruction is not placed opposite a door to a room if it would prevent a wheelchair user turning into or out of the room; and

c. doors to habitable rooms and a room containing a WC have minimum clear opening widths shown in Table 1, when accessed by corridors or passageways whose widths are in accordance with those listed in Table 1.

Table 1 shows the minimum widths of corridors and passageways that would be necessary to enable wheelchair users to turn into and out of a range of doorway widths.

Table 1 Minimum widths of corridors and passageways for a range of doorway widths

Doorway clear opening width (mm)	Corridor/passageway width (mm)
750 or wider	900 (when approach head-on)
750	1200 (when approach not head-on)
775	1050 (when approach not head-on)
800	900 (when approach not head-on)

Diagram 21 illustrates the guidance on doors, corridors and passageways.

Diagram 21 **Corridors, passages and internal doors**

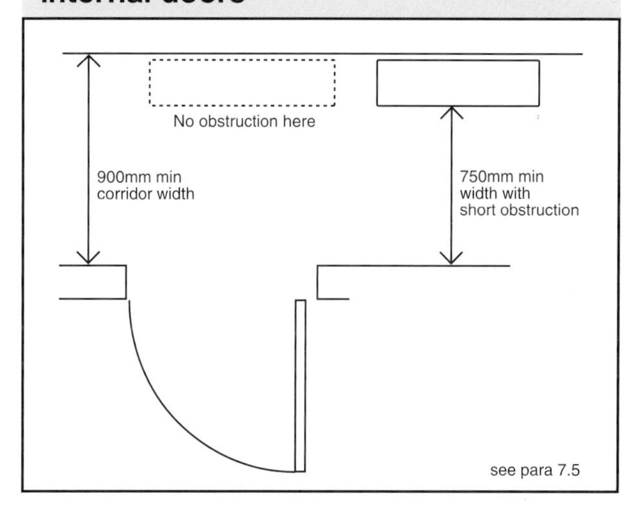

Vertical circulation within the entrance storey

Design considerations

7.6 In exceptional circumstances, where severely sloping plots are involved, a stepped change of level within the entrance storey may be unavoidable. In those instances, the aim should be to provide a stair of reasonable width for ambulant disabled people to negotiate the steps with assistance and for handrails on both sides. The Approved Document to Part K1 of the Building Regulations contains guidance on the design of private stairs in dwellings.

Provisions

7.7 A stair providing vertical circulation within the entrance storey of the dwelling will satisfy the Requirement if:

a. it has flights whose clear widths are at least 900mm;

b. there is a suitable continuous handrail on each side of the flight and any intermediate landings where the rise of the flight comprises three or more rises; and

c. the rise and going are in accordance with the guidance in the Approved Document for Part K for private stairs.

Section 8

ACCESSIBLE SWITCHES AND SOCKET OUTLETS IN THE DWELLING

Objective

8.1 The aim is to assist those people whose reach is limited, to use the dwelling more easily by locating wall-mounted switches and socket outlets at suitable heights.

Design considerations

8.2 Switches and socket outlets for lighting and other equipment should be located so that they are easily reachable.

Provisions

8.3 A way of satisfying the requirements would be to provide switches and socket outlets for lighting and other equipment in habitable rooms at appropriate heights between 450mm and 1200mm from finished floor level (see Diagram 22).

Diagram 22 **Heights of switches, sockets etc**

see para 8.3

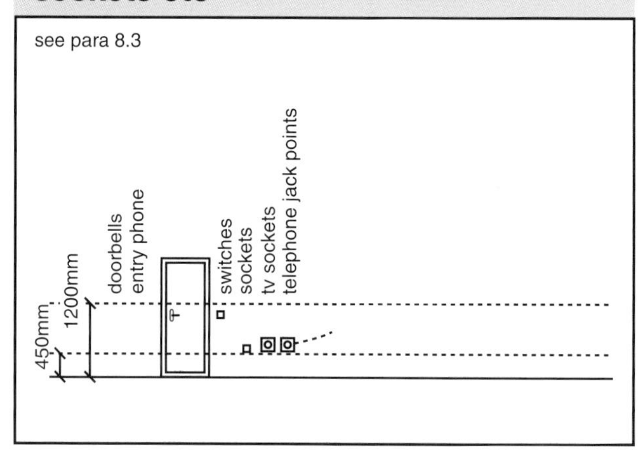

Section 9

PASSENGER LIFTS AND COMMON STAIRS IN BLOCKS OF FLATS

Objectives

9.1 For buildings containing flats, the objective should be to make reasonable provision for disabled people to visit occupants who live on any storey.

9.2 The most suitable means of access for disabled people from one storey to another is a passenger lift. However, a lift may not always be provided.

Design considerations

9.3 If there is no passenger lift providing access between storeys, a stair should be designed to suit the needs of ambulant disabled people. In any event, a stair in a common area, should be designed to be suitable for people with impaired sight.

9.4 Where a lift is provided, it should be suitable for an unaccompanied wheelchair user. Suitable provision should also be made for people with sensory impairments. Measures should also be adopted which give a disabled person sufficient time to enter the lift to reduce the risk of contact with closing doors.

Provisions for common stairs

9.5 The Requirement will be satisfied if a building containing flats, in which a passenger lift is not to be installed, is provided with a suitable stair, which has:

a. all step nosings distinguishable through contrasting brightness;

b. top and bottom landings whose lengths are in accordance with Part K1;

c. steps with suitable tread nosing profiles (see Diagram 23) and uniform rise of each step, which is not more than 170mm;

d. uniform going of each step, which is not less than 250mm, which for tapered treads should be measured at a point 270mm from the inside of the tread;

e. risers which are not open; and

f. a suitable continuous handrail on each side of flights and landings if the rise of the stair comprises two or more rises.

Diagram 23 **Common stairs in blocks of flats**

see para 9.5

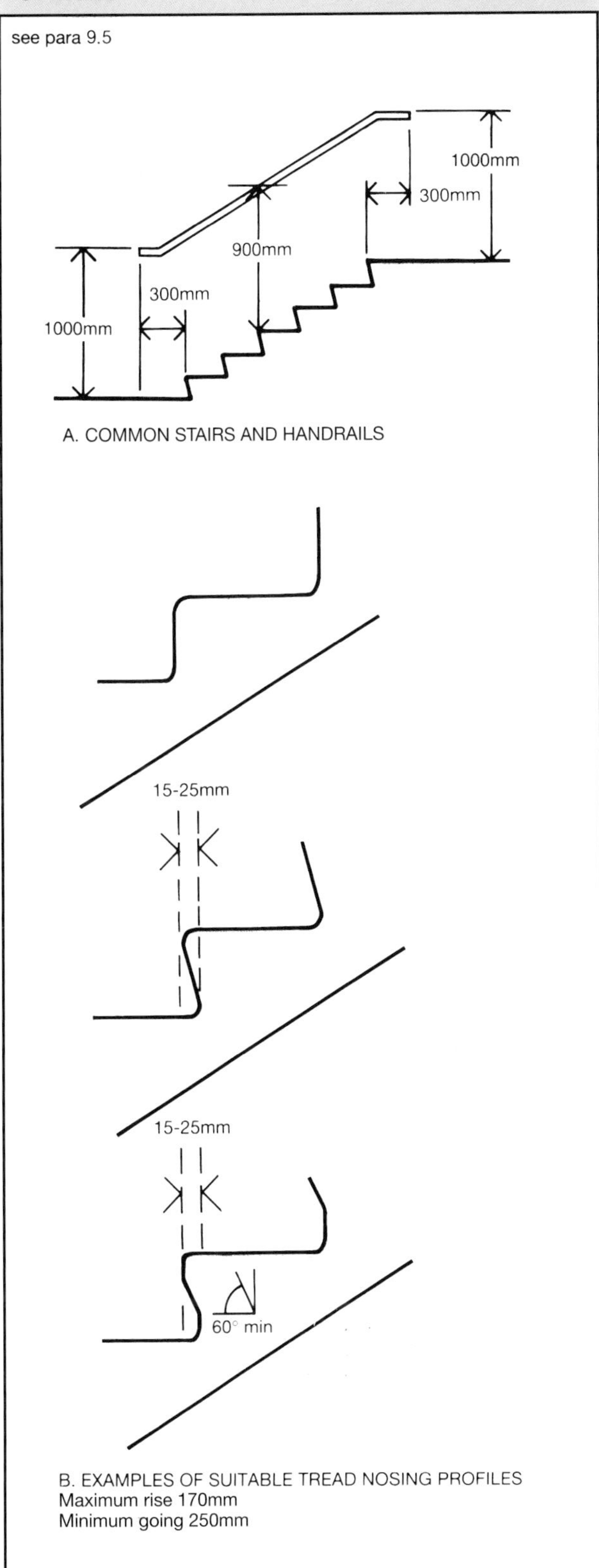

Provisions for lifts

9.6 The Requirement will be satisfied if a building, or a part of a building which contains flats above the entrance storey and in which passenger lift access is to be installed, is provided with a suitable passenger lift with a minimum load capacity of 400kg.

9.7 One way of satisfying the Requirement would be to provide a passenger lift which:

a. has a clear landing at least 1500mm wide and at least 1500mm long in front of its entrance;

b. has a door or doors which provide a clear opening width of at least 800mm;

c. has a car whose width is at least 900mm and whose length is at least 1250mm (other dimensions may satisfy the Requirement where shown by test evidence or experience in use, or otherwise, to be suitable for an unaccompanied wheelchair user);

d. has landing and car controls which are not less than 900mm and not more than 1200mm above the landing and the car floor, at a distance of at least 400mm from the front wall;

e. is accompanied by suitable tactile indication on the landing and adjacent to the lift call button to identify the storey in question;

f. has suitable tactile indication on or adjacent to lift buttons within the car to confirm the floor selected;

g. incorporates a signalling system which gives visual notification that the lift is answering a landing call and a 'dwell time' of five seconds before its doors begin to close after they are fully open: the system may be overriden by a door re-activating device which relies on appropriate electronic methods, but not a door edge pressure system, provided that the minimum time for a lift door to remain fully open is 3 seconds; and

h. when the lift serves more than 3 storeys, incorporates visual and audible indication of the floor reached.

Section 10

WC PROVISION IN THE ENTRANCE STOREY OF THE DWELLING

Objectives

10.1 The primary objective is to provide a WC in the entrance storey of the dwelling and to locate it so that there should be no need to negotiate a stair to reach it from the habitable rooms in that storey. Where the entrance storey contains no habitable rooms, it is reasonable to provide a WC in either the entrance storey or the principal storey.

Design considerations

10.2 The aim is to provide a WC within the entrance storey or the principal storey of a dwelling. Where there is a bathroom in that storey, the WC may be located in that bathroom. It will not always be practical for the wheelchair to be fully accommodated within the WC compartment.

Provision

10.3 The Requirement will be satisfied if:

a. a WC is provided in the entrance storey of a dwelling which contains a habitable room; or where the dwelling is such that there are no habitable rooms in the entrance storey, if a WC is provided in either the entrance storey or the principal storey;

b. the door to the WC compartment opens outwards, and is positioned to enable wheelchair users to access the WC and has a clear opening width in accordance with Table 1 (door openings wider than the minimum in Table 1 allow easier manoeuvring and access to the WC by wheelchair users); and

c. the WC compartment provides a clear space for wheelchair users to access the WC (see Diagrams 24 & 25) and the washbasin is positioned so that it does not impede access.

Diagram 24 **Clear space for frontal access to WC**

see para 10.3

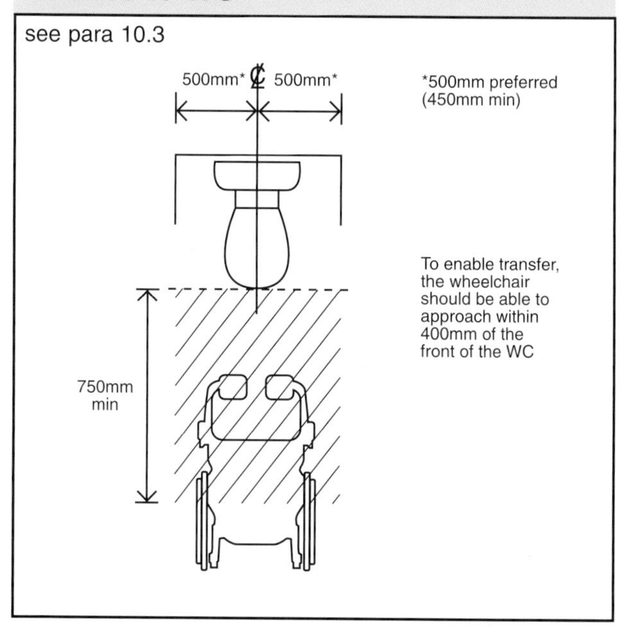

Diagram 25 **Clear space for oblique access to WC**

see para 10.3

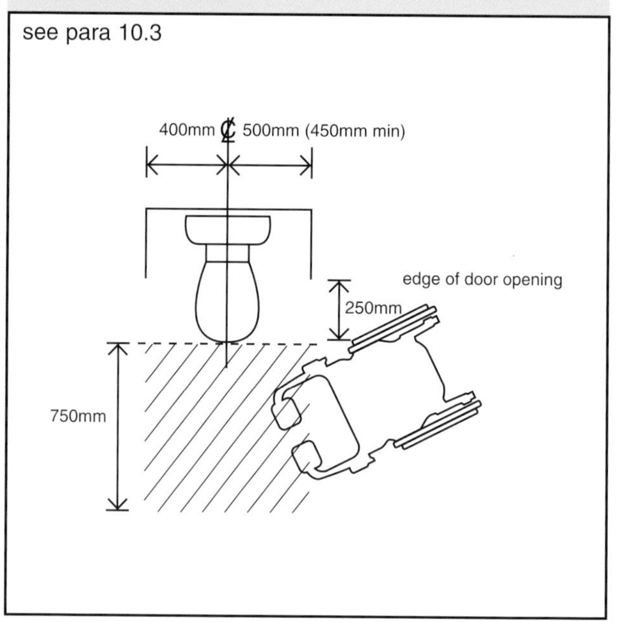

Standards referred to

BS 4787: *Internal and external wood doorsets, door leaves and frames:*
Part 1: 1980 (1985) *Specification for dimensional requirements*

BS 5655: *Lifts and service lifts:*
Part 1: 1986 *Safety rules for the construction and installation of electric lifts*
Amendment slip
1: AMD 5840
(Part 1 to be replaced by BS EN 81-1, when published)
Part 2: 1988 *Safety rules for the construction and installation of hydraulic lifts*
Amendment slip
1: AMD 6220
(Part 2 to be replaced by BS EN 81-2, when published)
Part 5: 1989 *Specifications for dimensions for standard lift arrangements*
Part 7: 1983 *Specification for manual control devices, indicators and additional fittings*
Amendment slip
1: AMD 4912

BS 5776: 1996 *Specification for powered stairlifts*

BS 6440: 1983 *Code of practice for powered lifting platforms for use by disabled persons*
(Amendment due 1999)

Other publications referred to

Guidance on the use of Tactile Paving Surfaces, Department of the Environment, Transport and the Regions, 1997.
Copies of this document may be obtained from:

The Mobility Unit
Department of the Environment, Transport and the Regions
Zone 1/11
Great Minster House
76 Marsham Street
LONDON SW1P 4DR

Guide to Safety at Sports Grounds
London: The Stationery Office, 1997

Designing for Spectators with Disabilities
Football Stadia Advisory Design Council, 1993
(available from the Sports Council)

Access for disabled people
English Sports Council Guidance Notes